SAVE THE ANIMALS!

POLAR BEARS

by Emma Huddleston

Consultant: Beth Gambro
Reading Specialist, Yorkville, Illinois

BEARPORT PUBLISHING

Minneapolis, Minnesota

Teaching Tips

Before Reading

- Look at the cover of the book. Discuss the picture and the title.
- Ask readers to brainstorm a list of what they already know about polar bears. What can they expect to see in this book?
- Go on a picture walk, looking through the pictures to discuss vocabulary and make predictions about the text.

During Reading

- Read for purpose. Encourage readers to think about how to help polar bears as they are reading.
- Ask readers to look for the details of the book. What are they learning about the problems polar bears are facing?
- If readers encounter an unknown word, ask them to look at the sounds in the word. Then, ask them to look at the rest of the page. Are there any clues to help them understand?

After Reading

- Encourage readers to pick a buddy and reread the book together.
- Ask readers to name one problem from the book polar bears face and one solution. Find the pages that tell about these things.
- Ask readers to write or draw something they learned about saving polar bears.

Credits:
Cover and title page, © HuntedDuck/iStock; 3, © ILYA AKINSHIN/Shutterstock; 5, © jo Crebbin/Shutterstock; 6–7, © Caleb Foster/Shutterstock; 8–9, © Joe McDonald/Shutterstock; 11, © HeliRy/iStock; 13, © james_stone76/Shutterstock; 14–15, © Keith Levit/Shutterstock; 17, © NPS Photo/Alamy; 18–19, © FatCamera/iStock; 21, © Anne Kiel/Shutterstock; 22, © Christopher Wood/Shutterstock; 23TL, © belterz/iStock; 23TC, © ansonmiao/iStock; 23TR, © Paolo Bona/Shutterstock; 23BL, © GTW/Shutterstock; and 23BR, © sandsun/iStock.

Library of Congress Cataloging-in-Publication Data

Names: Huddleston, Emma, author.
Title: Polar bears / by Emma Huddleston.
Description: Minneapolis, Minnesota : Bearport Publishing Company, [2023] | Series: Save the animals! | Includes bibliographical references and index.
Identifiers: LCCN 2022009808 (print) | LCCN 2022009809 (ebook) | ISBN 9781636917948 (library binding) | ISBN 9781636918013 (paperback) | ISBN 9781636918082 (ebook)
Subjects: LCSH: Polar bear--Juvenile literature. | Polar bear--Conservation--Juvenile literature.
Classification: LCC QL737.C27 H828 2023 (print) | LCC QL737.C27 (ebook) | DDC 599.786--dc23/eng/20220308
LC record available at https://lccn.loc.gov/2022009808
LC ebook record available at https://lccn.loc.gov/2022009809

For more information, write to Bearport Publishing, 5357 Penn Avenue South, Minneapolis, MN 55419. Printed in the United States of America.

Contents

On Thin Ice

Drip, drip, drip.

Arctic ice is **melting**.

But this is the polar bear's home.

What will happen to the big bear?

Polar bears are made for the cold.

They swim in chilly water and rest on **floating** ice.

The bears also find their food on the ice.

They **hunt** seals that live there.

But Earth is getting hotter every year.

People are a big part of the problem.

They put bad **gases** in the air.

This makes it warmer.

When it is warmer, the ice melts fast.

Then, there are not as many seals.

Polar bears must work harder to find food.

Some polar bears have lost their icy homes.

They have to live on land.

These bears sometimes eat trash.

This makes them sick.

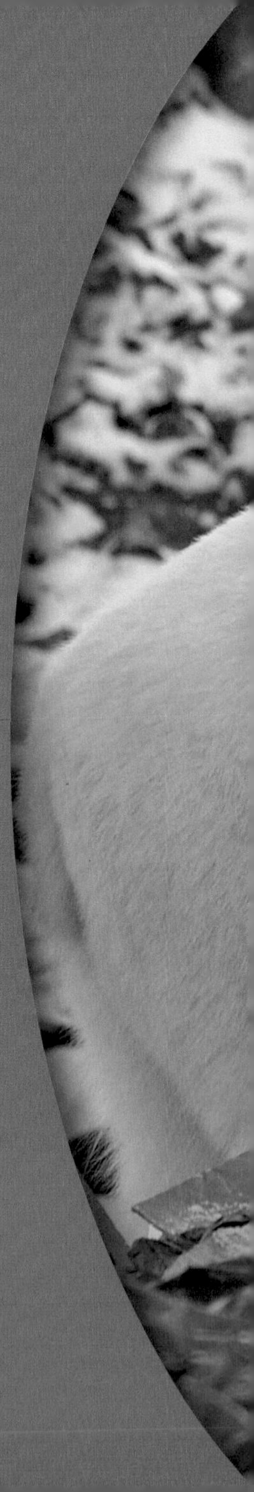

But many people want to help polar bears.

They are cleaning up the land.

Some are trying to stop Earth from getting hotter.

07189

You can help, too!

Driving cars makes bad gases.

Ride your bike or walk to get around.

Look at the polar bears playing.

Together, we can save the polar bears!

Polar Bears in the Wild

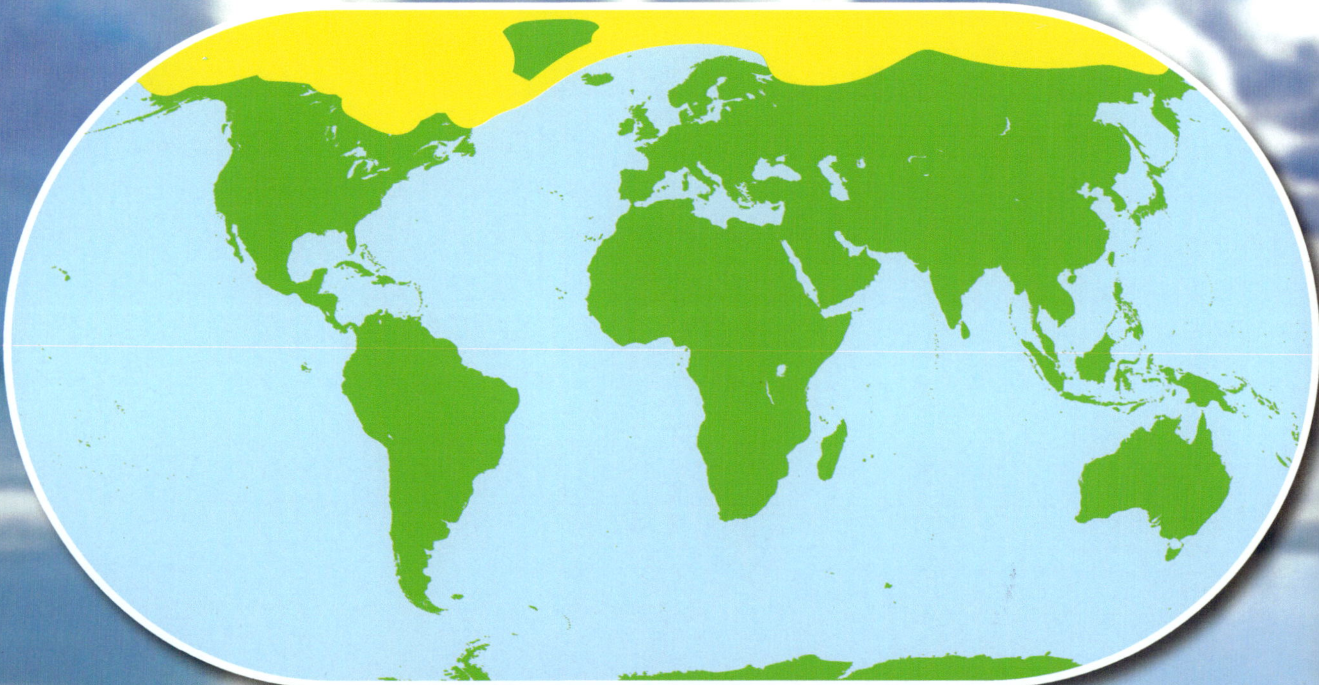

Where polar bears live

Glossary

Arctic the farthest north area on Earth

floating sitting on top of water

gases things that float in the air

hunt to look for other animals to eat

melting turning to liquid because of heat

Index

Read More

Hansen, Grace. *Help the Polar Bears (Little Activists: Endangered Species).* Minneapolis: Abdo Kids Jumbo, 2019.

Schuh, Mari C. *Polar Bears (The World of Ocean Animals).* Minneapolis: Jump!, Inc., 2022.

Learn More Online

1. Go to **www.factsurfer.com** or scan the QR code below.
2. Enter "**Save Polar Bears**" into the search box.
3. Click on the cover of this book to see a list of websites.

About the Author

Emma Huddleston lives in the Twin Cities with her husband and daughter. She enjoys writing children's books and spending time outside.